What is inside this book

- **Single Podcaster**
- **Startup Gear (Basic)**
- **Better Gear (Customize)**
- **Recording Software**
- **Hosting options**
- **Podcast Syndication**
- **Be You and Enjoy**
- **Podcast Mindset**
- **Getting Guests**
- **Build and Audience**
- **Bundle Option**
- **Software Recorders**
- **Software Editors**
- **Monetize**
- **Sponsors**
- **Phone to Computer**
- **Smart Phone to Phone**
- **Sample Your Audio**
- **Multi Mic setup**

From an image to reality

Being introduced to Anthony was the day my life changed forever. We met by sheer coincidence. I was on my way out of the building and he on the way in. We were introduced by Richie, a friend I was there visiting.

Richie had to leave and me and Anthony started talking. He was telling me about his vision to create the second level of that building into a full functioning studio. At this time, I could not understand what he was explaining to me because I've never been around the filming industry. He invited me to go upstairs for a tour, of what was once a large space cluttered with yard sale items and large furniture to its radical improvement today.

He began to paint a picture of this space being transformed into something I've only seen on tv. A greenroom with a host and guest interviewing table over here, this section will be for headshots, well do promotional videos in this area and I was starting to see the studio concept come to life.

Forbes Riley at the Icon Summit

IK Multimedia iRig PRO DUO Studio Suite

- ake two channels of 24-bit audio on the road with the iRig PRO DUO

- Capture breathtaking performances with the iRig Mic Studio XLR

- Monitor yourself accurately with the closed-back iRig Headphones

Smart Phones can be used for many podcasting productions using Audio apps

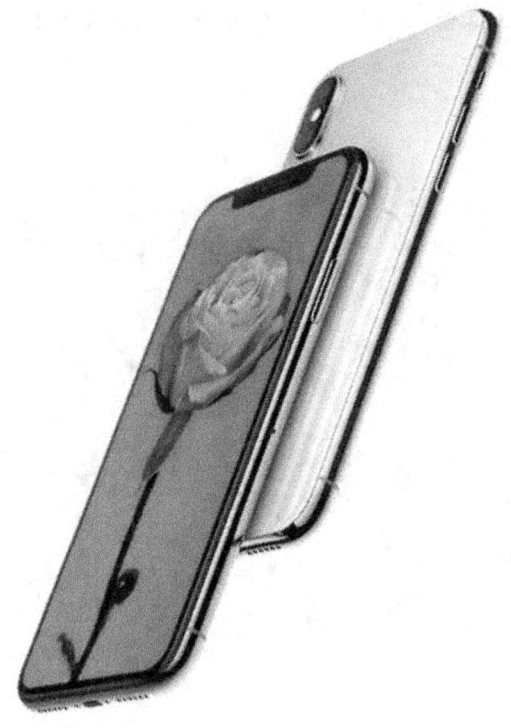

Many of the Options in this book are designed for multiple platforms with great results

Presonus Audiobox USB DAW Recording Bundle with Studio One Artist Recording Software and 10ft LYX LCS Premium XLR Cable

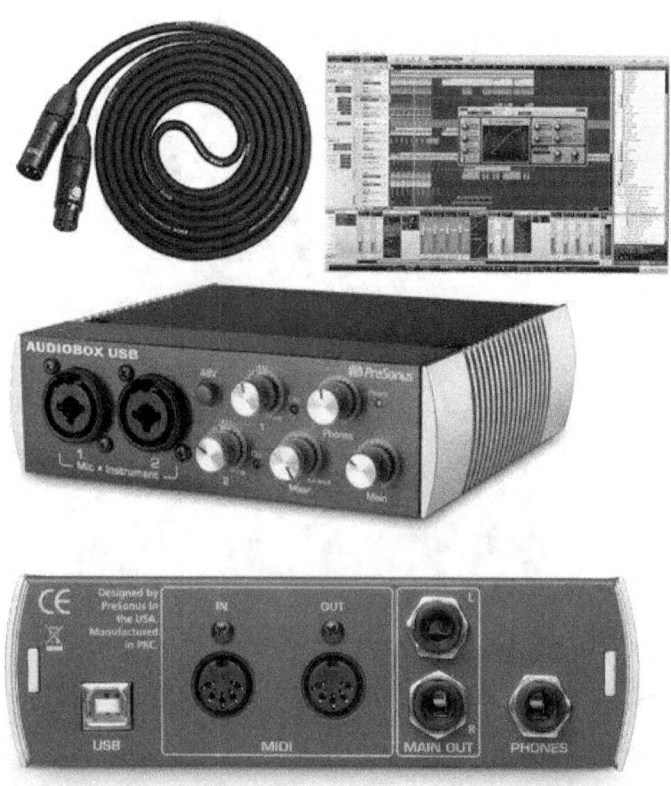

Studio One Artist Recording Software

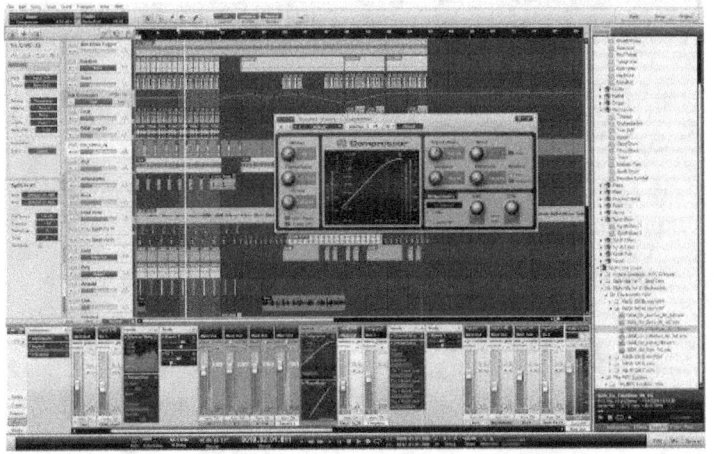

Fully working demo Software With the ability to make audio sound like studio quality from almost any space

http://studioone.presonus.com

Producing interviews and podcasts at near studio-quality like audio directly to your Smartphone using Low-cost high-quality options. Several package options are convenient and convenient enough for an individual to manage the entire production.

Background noise is a plus when recording interviews on location it Ads to the value of your interview, but it should not distract from the interview itself.

VOCALIVE FOR THE IPAD GIVES VOCALISTS MORE TO PLAY WITH

Your favorite vocal effects now on your iPad

Great Options built in, Compressor, Phazer Chorus, Parametric EQ, choir, Delay, Env Filter, Double, Reverb, Morph and more.

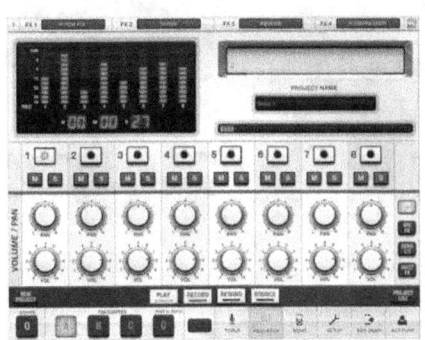

Core i5 or i7 or 4 Core equivalent at the minimum dedicated computer that is tuned with all your needed Software or Apps

A Dedicated computer is essential for error-free podcasting. The 2nd computer for live streaming is best.

Microsoft LifeChat LX-3000 Headset

- **Premium stereo sound**
- **Excellent reliability and clarity**
- **Noise canceling microphone**
- **Comfortable leatherette ear pads**
- **Optimized for Skype**

- **Clarity, Comfort, and Convenience - Digital USB for superior clarity, built-in unidirectional microphone with acoustic noise cancellation**
- **Leatherette ear pads for improved comfort, in-line volume controls**
- **Pivoting boom microphone with 180-degree movement, flexible 6-foot cable**

Tablet PC's i5 or i7 4 core with 4 or 8 logical processors for best results

The Mobile Podcaster

Mobile Journalism with Shure MOTIV™ MVL Mobile Lavalier Microphone

PLUG IN, CLIP ON, SPEAK OUT.

MVL OMNIDIRECTIONAL LAVALIER MICROPHONE FOR MOBILE DEVICESimply clip the MVL omnidirectional condenser lavalier microphone to your collar or lapel and plug it into your mobile device to capture clear, quality audio for interviews, lectures, public speaking and videography. You can edit your recordings with the free ShurePlus™ MOTIV™ iOS app.

Finally, spread the word — send your sound files via text, email or any file sharing service.

- For use with iOS and Android[*] devices with recording capabilities
- Compatibility with the ShurePlus™ MOTIV™ app and other iOS audio and video recording apps
- Plugs into headphone jack via a 1/8" (3.5mm) TRRS connector
- Omnidirectional microphone with exceptional signal-to-noise ratio
- Includes windscreen, clothing clip, and carrying pouch

iPhone Microphone
iPad Microphone
iPod Microphone
Android Microphone[*]
USB Microphone

Condenser Microphone for iOS and USB The MV5 offers professional-quality audio, with the flexibility and control of switchable DSP recording presets and latency-free headphone monitoring.

NOW EVERY ROOM IS A VOICEOVER ROOM.

MV5 CONDENSER MICROPHONE FOR iPHONE, iPAD & USB

With an iconic design that inspires vocal performance, the MV5 digital condenser microphone offers professional-quality audio, plus the flexibility of both iOS and USB connectivity.

Plug it in and get to work right away, or take advantage of three onboard DSP presets to quickly dial in the right sound for your project.

Zero-latency headphone monitoring with volume and mute, and an adjustable stand completes the recording experience.

- 2-in-1 iOS and USB connectivity offers instant setup at home or on the go
- Zero-latency headphone output with mute and volume control
- Compatible with the free ShurePlus MOTIV iOS app and other iOS audio/video apps

- Tuned to capture the human voice, time after time
- Lightning® connector and USB cable included

iPhone Microphone
iPad Microphone
iPod Microphone
Android Microphone[*]
USB Microphone

SENSATIONAL STEREO SOUND, SIMPLIFIED.

SENSATIONAL STEREO SOUND, SIMPLIFIED.
MV88 STEREO CONDENSER MICROPHONE FOR iPHONE & iPAD

The MV88 iOS digital stereo condenser microphone offers vloggers unrivaled convenience and professional-quality audio on the go.

Simply plug it into your iPhone, iPad, or iPod and hit record in your favorite audio or video recording app to get life-like stereo recordings. Or, personalize your sound with enhanced control over stereo width, polar patterns, EQ, and more with the free ShurePlus MOTIV app. Create, capture, and share your way, anywhere.

- Record, edit, and share recordings from the free ShurePlus MOTIV mobile app

- Lightning™ connector offers quick setup, plugging directly into an iPhone, iPad, and iPod
- Compatibility with leading iOS audio and video recording apps
- Microphone pivots and rotates, allowing you to establish the most ideal direction for recording
- Adjustable stereo width. Plus, choose from mid-side stereo, bidirectional, or cardioid only
-

iPhone Microphone
iPad Microphone
iPod Microphone

Camera Mount Wireless

Sennheiser AVX Camera Mountable Combo Wireless Set, Includes Handheld Transmitter, Bodypack Transmitter, Plug-On Receiver, ME 2 Lavalier Microphone, 1880-1930MHz

Camera Mount Wireless

Samson Concert 88 Camera Combo UHF Wireless System

Concert 88 Camera Combo UHF Wireless System from Samson features a 300' wireless range and 16 selectable frequencies for interference-free operation.

Tampa Mayor Dick Greco
Local News is Now wide open for
local community Podcasting

The Single Person Show Recordings

This setup is perfect for the traveler podcaster Smartphone or Tablet recoding The iRig Mic tabletop stand with your choice of Tablet and BossJock APP for IOS.

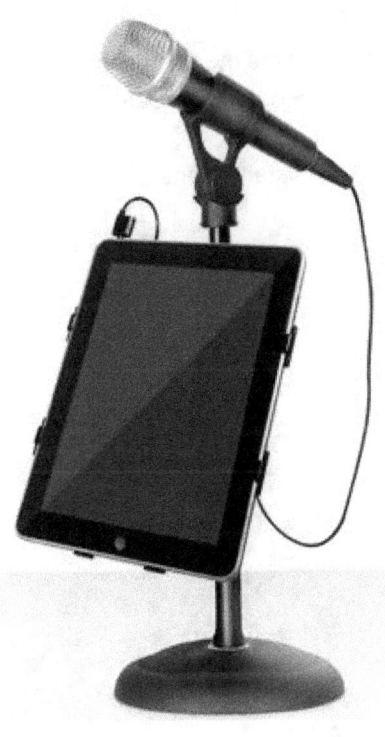

IK Multimedia iRig PRE for iPhone/iPod touch/iPad and Android Devices

- Works with all popular audio & video apps
- 30 hours battery life / 10 hours with phantom power on
- 15.75" cord makes it easy to connect and position
- +48v Phantom power for use with professional condenser microphones
- 1/8" headphone output for real-time monitoring with supported apps1/8" headphone output for real-time monitoring with supported apps
- XLR microphone input with adjustable sensitivity

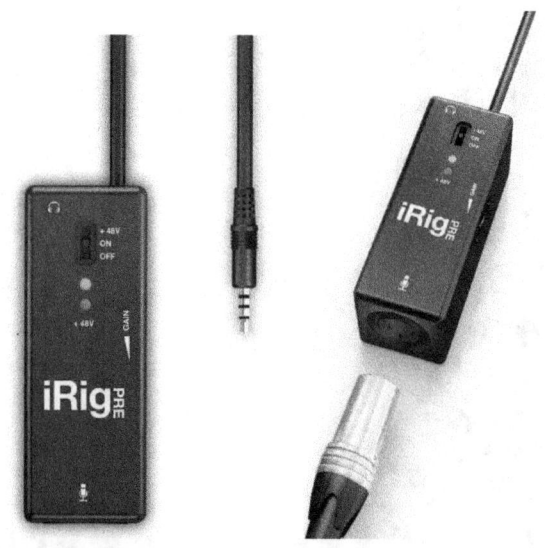

Samson R21 Dynamic Vocal Microphone

- High Output Dynamic Element
- Unidirectional Cardioid Polar Pattern for Maximum Gain Before Feedback
- Withstands High Sound Pressure Levels
- Dual Stage Windscreen
- Excellent for Live Performance and Recording

Shure SM58-CN Cardioid Dynamic Vocal Microphone

You'll find multiple SM58s in most any studio or live music venue - no doubt you've heard of them. The SM58's frequency response is tailored for vocals, with brightened midrange and bass roll off. The uniform cardioid pickup pattern isolates the main sound source and minimizes background noise, which is ideal for vocals, especially in live situations where background noise is substantial.

The microphone's pneumatic shock-mount system cuts down handling noise, making it a great choice for handheld vocals, and the built-in wind/pop filter helps keep things in check.

XLR Cable 3 or 6 Foot

- XLR connectors with internal strain relief for
    rugged reliability
- Oxygen-Free Copper (OFC) conductors
    for enhanced signal clarity
- Hi-density OFC braided shield for superior
    EMI and RFI rejection

IK Multimedia iRig Mic Studio Digital Studio Microphone

iRig Mic Studio is the latest addition to IK's high-quality digital microphone line. It packs a 1 inch. diameter condenser capsule into an ultra-compact enclosure that can be used with iOS (iPhone iPad iPod touch Mac), PC and Android devices. Roughly the same length as an iPhone and with a diameter of just 45mm iRig Mic Studio is less than half the size of competing large-diaphragm microphones. iRig Mic Studio's large-diaphragm cardioid electret condenser capsule 133dB SPL rating 24-bit 44.1/48kHz converter high-quality low-noise preamp gain control knob multi-color LED level indicator sturdy tripod base and other professional features ensure superior sound for every session. iRig Mic Studio also includes powerful apps and software: experience superior vocal processing with Voca Live unparalleled mobile editing

and recording with iRig Recorder and a vast collection of amps and stomp box effects with Amplitude. With iRig Mic Studio musicians vocalists home producers and more have a studio-quality large-diaphragm microphone that fits in the palm of their hand

- Professional studio microphone with large-diaphragm capsule

- "High-quality 1" back electret condenser capsule 24-bit converter with 44.1/48Khz sampling rate"

- Low-noise, high-definition preamp

- Integrated headphone output

- Onboard gain control and headphone level control

Over the Mic Windscreen

Most mics have a Build in windscreen but for those outdoor windy days best to have one available

SwissGear Mono Sling Gear Bag

Keep your gear at the ready but sling it behind you when you are doing interviews Helps to have your hands free when interviewing

BossJock APP for IOS.

Podcasters, Voiceover Artists, DJs, this is your app. Life without fan noise is good!

http://bossjockstudio.com

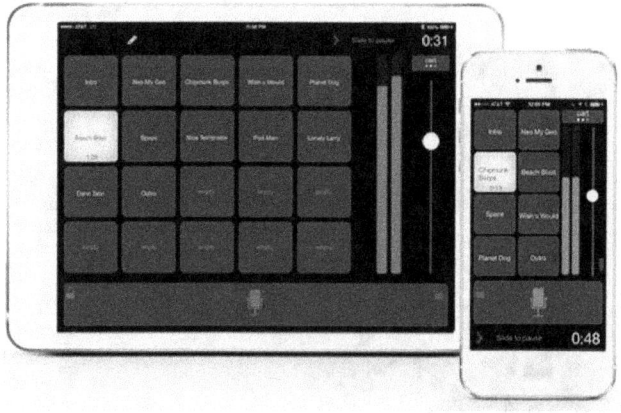

- Record a podcast while triggering intros/outros, bumpers and background music all on-the-fly Audiobus support: Receive live audio directly to bossjock studio from other Audiobus-compatible apps and stream live audio directly to other Audiobus-compatible apps!

- Dynamics processing (Compression/Limiting) is applied to the mic and mix for loud, level audio

- Up to 35 audio carts duck and fade behind the mic for dynamic sound and smooth transitions

- Audio is Podcast ready, no need for post-production

- Load carts with audio from your Music Library, Dropbox, Email, Wifi or AudioPaste from other music apps

- Fine-tune Cart playback in and out points as well as Cart color and behaviors including looping, fades, auto-rewind and ducking

- Send your recordings out via FTP, Dropbox, Soundcloud, iTunes, Wifi, iTunes Share, AudioCopy and Email

- Encodes to all your favorite formats – mp3, m4a, wav, aiff

Works with external mics like the Apogee Mic, IK Multimedia iRig Mic Cast and Blue Mikey

Full VoiceOver compatibility for visually impaired producers

Optimized for every iOS device. Perfect for the iPad Mini!

Turnkey publishing for Podcasters, Audioboo and Soundcloud Producers, Voice Over artists, Audio Bloggers and Mobile Broadcasters

Audio-Technica ATH-M40x
Professional Monitor Headphones

- **Cutting-edge engineering and robust construction**

- **40 mm drivers with rare earth magnets and copper-clad aluminum wire voice coils**

- **Tuned flat for incredibly accurate sound monitoring across the entire frequency range**

- **Circumaural design contours around the ears for excellent sound isolation in loud**

-

- **environments**

Monitor Speakers

PreSonus Eris E4.5 2-Way Powered Studio Monitors (Pair)

You must be able to listen to your production with clean sound monitors

The better the sound at edit the better it will sound on the consumer listener end.

The iRig Mic
For iPhone or Samsung
Smartphones is the simples option
but No control of the Type of Mic

Simple on one interviews at any
occasion by using your Smartphone
with the **IK Multimedia iRig PRE
Samson Mic and 3 foot cable**

Takes 30 seconds to plug in the
options but if you have the sling bag
as show you can have all things
ready in 12 seconds it takes to
access your smartphone app
BossJock the app is always ready
with your setting intact just start
recording

Insignia™ - Handheld Reporter Style Microphone

Ensure clear voice audio for video posts with this reporter-style Insignia handheld microphone. Its premium mic capsule optimizes voice recording quality, while its metal housing enhances noise isolation and provides durability. This Insignia handheld microphone connects to 3.5mm inputs and includes an adapter kit to support smartphone and tablet connectivity.

IK Multimedia iRig PRE
Give you complete control of the type of Mic for your voice

Your Mic XLR Mic Options or Endless but recommend the Samson R21 at about the

Hosting Podcast RSS

How does it work?
RSS FEED
What do you need from them?
They provide the RSS Feed Link for
your account

What do you do with the RSS Link?
You give it to the Syndication sites

Note: it is best to purchase the microphone that best fits your voice for your best results

We use some of the same equipment for the TV and Video production needs

Studio One Artist
Recording Software

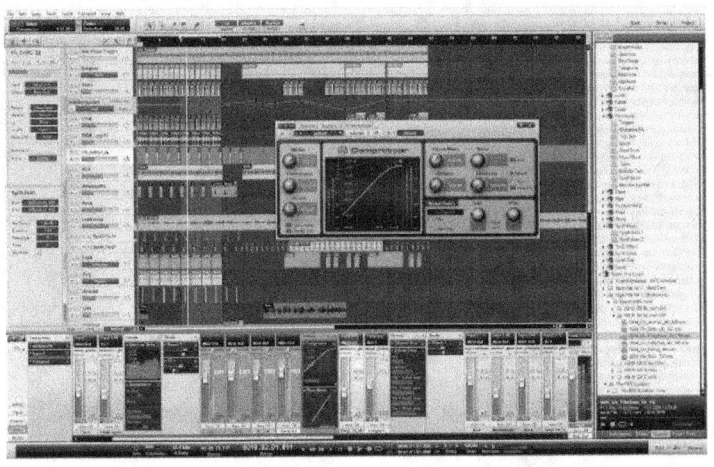

Fully working demo Software

http://studioone.presonus.com

Shure X2U XLR-to-USB Signal Adapter

- Plug and Play USB Connectivity allows the convenience of digital recording
- Provides +48V Phantom power for use with condenser microphones
- Includes Padded, zippered pouch and USB cable (3m / 9.8ft)
- Monitor Mix Control for blending microphone and playback audio
- Integrated pre-amp with Microphone Gain Control allows control of input signal strength

Insignia™ - Omnidirectional Lapel Microphone

Capture crisp, clear vocals with help from this Insignia™ NS-DLMIC10P lapel microphone, which features a high-quality microphone capsule and omnidirectional pickup pattern for optimal performance.

I-tunes
Podcast Syndication

All they require is to have an account with iTunes and submission of the RSS FEED

Continued ...... After a few months I came back to the location to see Anthony was upstairs painting the green room, the mess was cleared, and the studio was almost ready to produce videos, record commercials and setup for podcasting. Now it was time to get me ready for the camera.

He had a challenging road ahead with me. The cameras came on, lights in my face and I had that deer in the headlights look, it was terrifying, and I became paralyzed. Anthony's experience and knowledge of being in the media arena for many years he guided me through the fears that rattled through my mind about speaking in front of a camera. He was preparing me for what was about to happen next.

David Fagan

"Adele is a truth-teller.
More than knowledge she has
wisdom. She shares through her
brilliant questions and her humble
approach to life. Stop and listen and
I promise you'll be better for it.
David T Fagan"

I was introduced to celebrities, movie producers, podcasters, keynote speakers and way more. It all started with Podfest in Orlando, Fl. This was a 3-day event to bring Podcasters from all over the world together. This was new to me, I had no idea what podcasting was. The Podcasters would share their stories on how they got into podcasting.

How they made mistakes and gave tips on how to fix them. It was mind-blowing for me, I've never seen anything like it. Podfest is where I received my first taste of being introduced to a massive group people who were changing lives with their accomplishments.

I met a number of different people from all over the world. I was mesmerized and didn't want it to end, however, when it was over I took away experiences that I learned and grew from, to last a lifetime.

Google Play
Podcast Syndication

Podcasts, playing soon

Early Publishing Link
https://play.google.com/music/podcasts/publish

Podcasts are coming soon to Google Play Music. Add your shows now to connect with your audience and reach millions of potential listeners on Android.

Hosting of Podcast
Libsyn Libsyn.com

Offers an entry-level hosting for your podcast, simple to use

Libsyn provides everything your podcast needs: publishing tools, media hosting and delivery, RSS for iTunes, a Web Site, Stats, Advertising Programs, Premium Content, Apps for Apple, Android & Windows devices.

Buzzsprout
buzzsprout.com

Offers an entry-level hosting for your podcast, simple to use

Spreaker.com

Record a podcast or broadcast live using apps available for mobile, desktop, and the web. Plus upload files and migrate content over using the RSS Importer.

Distribute your podcast to social networks, iTunes, YouTube, and more. You can also schedule episodes, embed widgets, and order your own Mobile App.

Basic analytics help you measure your popularity through play numbers. Going Pro unlocks more details like sources, geolocation, and demographics.

Getting Guests

Building your talent for interviewing is more important than big-name guests start with local community personalities in your area. As your interview skills become refined.

Public Relations Firms

You can place yourself on P.R. organizations lists for specialty talent within your shows parameters that you can choose to schedule an interview with.

Eclectic Media PR Firm that manages all the interview opportunities Mediaproductions.tv

Social Media

Your communications via social media will be your simplest and best way to connect with your audience

Facebook, Linkedin, Google Plus just to name a few to develop a deep relationship with your audience. Letting your audience advocate their love for the show by setting up your posts to be easily shared.

Provide links to your blog posts, RSS FEED, Web Links with you shows organized and searchable.

Zoom for Facebook to record your one on one or multi interview podcast

Single Sign-On with Facebook OAuth

- **Features a simple and quick log-in, no need to create a password**
- **User can log-in to Zoom with an existing Facebook account**

Facebook IM: Presence

- **Zoom has integrated with Facebook IM**
- **See when your friends come online or go offline**
- **Search by name for any of your Facebook contacts**

Facebook IM: Instant Message

- **Instant message and start an instant meeting with your Facebook contacts**

Facebook: Video

- **Meet with up to 25 of your Facebook contacts**
- **Screen share documents, pictures and websites on your desktop, tablet or smartphone**

Google Hangouts
A Great way to Connect
with your interviews

Ringr App
A Great Tool for
Recording interviews
on your phone at Great
Quality

It works by recording both conversations at the phone with the app and both conversations get uploaded to the Cloud and stitched together as one interview as well as both conversations separately as well.

Monetize your Podcast
The plan should be in place of the purpose of your show (Podcast)

Is it a business you're already in or is it a hobby you enjoy?

When it is an existing business what you're doing is building leads for that business that connects you to a broader audience. That audience can become advocates for you and your business or directly convert to clients

Hobby or Niche you enjoy with extreme passion that builds an audience that audience related sponsors see you out in the new podcast media with direct downloads to your smart devices that carry over to your car for your listening enjoyment even a small audience in a specific niche has sponsor value to the appropriate advertiser who wants to fine-tune their marketing Dollars.

Sponsors

The marketplace for sponsors who are actively looking for specific audiences' that have a great deal of interest in their product that support their business, hobby, fun, work, learning and more. These sponsors know that it is more cost effective to be a sponsor or advertiser in a small audience that directly relates to their product than the mass-market promotion

They are looking for you and most likely are listening to your show (podcast) provide the great content and build your social audience.

Talk about the products you enjoy use why you use them that are relevant to the show content.

Choosing your format

Once you start do not stop if you stop you will lose all momentum.

1. Choose a monthly, weekly daily
2. Time of day you will upload
3. Length of time the show will be
4. In studio format or Mobile show
5. Get gear for specific type of show optimized and at the ready production dedicated gear.
6. Practice on friends and business associates to get the system and quality of production at a professional level. (You will improve as you produce each show. Listen to them for errors until you master the production and interview process)

Sample Your Audio

Each Mic and recording devices have their unique settings and in editing, you have an endless array of compression options for your voice.

1. Do you want your equipment to be the vocal quality you want?
2. Do you want your editing to fix the sound such as base with so many options you need to sample your sound from the final product?
3. Understanding how to replicate your settings so they can be reproduced flawlessly
4. If you're hiring an Audio Professional manage your production make sure you have the settings saved and the software used so it can be replicated if you're using special customization settings.

Skype
http://www.skype.com
Internet phone calls

Skype to Skype is a free service and to make a Skype to phone requires a billing plan

Several Auto Skype recorders that will record your Skype Recording Easy to use one-click record

Best quality is the Skype to Skype conversation rather than Skype to phone

The quality compression on phones are very high which when recorded is incredibly low compared to your available recording options.

http://www.evaer.com
Evaer is an add-on video recorder for Skype. You may freely use the trial-version of Evaer Video Recorder for Skype and can record only up to 5 minutes for each video calls. Buy the following license you can remove the recording limitation

Celebrity Connections

As you build your fan base your guests will become celebrities and your access to established celebrity guests will become easier regardless of the niche it is the quality of the fan base and production over time with better equipment, technique, and style.

After time you can consider publishing books, videos to reach other markets with your podcast.

The Niche Celebrity

A celebrity that is respected based on work done and the number of people they communicate to on consistent basis that makes them an authority in a specific space that can carry over to other levels of authority

Several of the people here at the time of this photo were unknown one has reach national success others or niche successful and expanding.

Photo "Anthony" while a rerun is in play and a one of the business briefs segments was on the other came out of the live show for a photo opportunity as 2 of my shows were running at the same time on 2 separate channels

Mike Alstott for a special tour of the
Raymond games stadium

**Phone rings on a live broadcast
Debate** "always turn off your phone"

Podcasting can become the opportunity you have been waiting for to communicate your ideas in any niche even a tiny niche can become successful.

Keven Herington of shark tank

Keven Herington of shark tank

Promoter investor. At an investor pitch event at the 36,000-square foot Estate in Tampa had some great advice on how to be ready for talking to an Angel investor for your product or service. You only have moments so have your numbers ready on how the investor benefits.

Modern Communication requires On Camera on Mic Experience

You must be able to provide guidance to your clients you will interview to get them to feel confident

Keep your tech as simple as possible so to get the content in any situation. People are on the move like never before a quick moment is the opportunity.

Podfest 2017

Continued....... By this time my head was spinning with amazement. I was grateful for this opportunity and intrigued about the future. Next huge event was the Icon Summit in Tampa, which we attended months later. David Fagan 'The Hollywood Entrepreneur" was the hosting this event. David has a

Real Estate Talk with Adele D'Argenio

Danny Collins w.EnergySolutionsDirect.com 352.601.6475

grand list of achievements, he sold over 23 million books, been on the Today Show with Matt Lauer, featured on Fox and Friends the list goes on. Wow, I got to meet him he is an incredible person and soon after we became great friends. At that event I also met Forbes Riley an internationally known award-winning TV host, author, keynote speaker and I had the pleasure to interview her. If it wasn't for Anthony pushing me out of my comfort zone, I never would have approached her myself.

Forbes has a unique way of changing people's lives. In the future I was invited by Forbes to attend an exclusive event at her home for three days. Who could ask for more. Oh yes, there's more.

The studio was ready for a talk show, with a host interviewing the guest setup. We had already done promotional videos for community events, business, as well as commercials. Valerie becomes the host of Town Talk and I was there to help. I was given the role of production assistant and with my quick learning skills eventually production manager. I was taught how to position the cameras, where to place the microphone on the host and guest, how lighting is critical for clarity. Learning how to set the volume for each person, there is a lot to know. The more I became trained with Anthony's direction, the better I was getting. I was comfortable with the operations behind the camera and ready for the next step, well not exactly.

Now, since Anthony was notorious for pushing me out of me comfort zone, it was time create a show for me. I was petrified of the idea. You know the old adage if something scares you, move towards it. Of course, not a good plan if it's a tiger. It was either do it now or regret later what I should have done. So, I jumped and with Anthony's guidance how could I fail. I made mistake, after mistake after redo, bumbled my words, lost my focus over and over again. It was challenging. I thought to myself "this is tough, I don't know what I'm doing, how did I get here" you know the doubts that pass through our minds from time to time. I still get a little nervous but not as scary as it was for my first show.

During this time, I was learning how to build a successful show. Preparing my guests for what to expect and setting up the time for them to be there for the recording. I would organize a list of questions to place in front of me because once the cameras came on I forgot everything.

I also had to prepare myself mentally because thousands of people would be viewing the show.

No matter how prepared you get shooting your first show is hard.

Because of Anthony's support and encouragement, he made the learning process enjoyable. What I've learned is having an extraordinary mentor completely dedicated to my success and growth, gave me the confidence and capability to continue. To me that's priceless and I'm curious to see what the future is holding for me!

The list of the interviews Anthony has been thanks to Shannon Rose, owner of Eclectic Media PR Firm that manages all the interview opportunities Mediaproductions.tv

WDUN 102.9 FM/550 AM The Ken Coleman Show Gainsville, GA The Schilling Show "Where the News is Made!"™ Newsradio 1070 WINA Charlottesville, VA WENG AM & FM Englewood, FL The Answer 1420 AM The Advocate w/ Nick Phillips Cleveland, OH Morning Update with Paul Linnman 1190 KEX Portland, OR SiriusXM - Channel 110 Radio Disney Detroit AM 910 Glen Biegel Show AM 700 KBYR Anchorage, AK 99501 WWNN 1470AM Boca Raton, FL WTVN Columbus, OH FM NewsRadio 106.9 WSYR Syracuse, NY 92.5 Fox News Ft. Myer, FL he Rude Awakening Show WOCM-FM Ocean City, MD Full Power Radio News Now 94.9 WJJF WSVA Radio Harrisonburg, VA Full Power Radio News Now 94.9 WJJF Price of Business KTEK 1110 AM Houston, TX The Shannon Burke Show Nationally Syndicated Talk of the Town WEZO 1230 Augusta, GA Talk of the Town WEZO 1230 Augusta, GA The Josh Tolley Show Nationally Syndicated Holistic Survival w/ Jason Hartman Nationally Syndicated Health, Wealth & Wisdom 1470AM WMGG Holistic Survival w/ Jason Hartman Atlanta, GA WYAY (106.7 FM) Austin, TX KJCE (1370 AM) Bend, OR KBNW (1340 AM/104.5 FM) Birmingham, AL WERC (105.5 FM) Birmingham, AL WYDE (101.1 FM) Chapel Hill, NC

WCHL (97.9 FM) Charleston, SC WTMA (1250 AM) Charlotte, NC WBT (1110 AM) Cleveland, OH WTAM (1100 AM) Cleveland, OH WHK (1420 AM)

Scott Goss

Cleveland, OH WHKW (1220 AM) Columbus, OH WTVN (610 AM) Columbus, OH WRFD (880 AM) Columbus, OH WJKR (103.9 FM) Denver, CO KLZ (560 AM) Denver, CO KNUS (710 AM) Des Moines, IA KWQW (98.3 FM) Des Moines, IA KRNT (1350 AM) Des Moines, IA KPSZ (940 AM) Detroit, MI WLQV (1500 AM) Fayetteville, AR KURM (790 AM/100.3 FM)
Fayetteville, NC WFNC (640 AM) Fresno, CA KMJ (580 AM/105.9 FM) Ft. Wayne, IN WOWO (1190 AM/92.3 FM) Greensboro, NC WPTI (94.5 FM) Greensboro, NC WTRU (830 AM/97.7 FM) Greensboro, NC WTIB (103.7 FM) Houston, TX KROI

Joseph Warren

(92.1 FM) Jacksonville, FL WOKV (690 AM) Las
Vegas, NV KXNT (100.5 FM) Louisville, KY WHAS
(840 AM) Loveland, CO KCOL (600 AM) Medford, OR
KMED (1440 AM) Minneapolis, MN WWTC (1280 AM)
Minneapolis, MN KKMS (980 AM) Minneapolis, MN
WCTS (1030 AM) Nashville, TN WLAC (1510 AM)
Orlando, FL WFLF (104.5 FM) Phoenix, AZ KTAR
(92.3 FM) Portland, OR KPAM (860 AM) Portland,
OR KXL (101.1 FM) Portland, OR KPDQ (800 AM)
Providence, RI WHJJ (920 AM) Raleigh, NC WTRU
(1030 AM) Raleigh, NC WTKK (106.1 FM) Richmond,
VA WRVA (1140 AM) Roanoke, VA WFIR (960
AM/107.3 FM) San Antonio, TX KTSA (550 AM) San
Francisco, CA KNEW (960 AM) San Francisco, CA
KKSF (910 AM) San Francisco, CA KFAX (1100 AM)

San Francisco, CA KDOW (1220 AM) Scranton, PA WILK (910 AM/103.1 FM) Southern Pines, NC WEEB (990 AM) Tampa, FL WFLA (970 AM) Tampa, FL WWBA (820 AM) Virginia Beach, VA WNIS (790 AM) Wenatchee, WA KPQ (560 AM) Wenatchee, WA KOZI (93.5 FM) Winston-Salem, NC WSJS (600 AM) York, PA WSBA (910 AM) TV Stations Money Business Life Network National TV Show: Boomers'

Santa

Braintrust Host: Johnny Dean State City Call Letters Station Alabama Huntsville/Decatur/Florence WZDX 54.1 Alabama Mobile WPMI NBC 15 Alaska Anchorage KTVA CBS11 Arizona Phoenix KTVK IND 3 Arizona Yuma KYMA NBC 41.1 Arkansas Fort Smith/Fayetteville/Sprindale/Rogers KNWA NBC 51.1 FOX 51.2 Arkansas Little Rock/Pine Bluff KLRT FOX 16 California Eureka KIEM NBC 3.1 California San Francisco KTVU 2.1 Florida Augusta WRDW-DT MNTV 12.2 Florida Gainesville WNBW NBC 9 Florida

Orlando WKCF CW 18 Florida Pensacola WPMI NBC 15 Florida Tallahasee WTWC NBC 40.1/40.2 Georgia Columbus WLTZ NBC 38 Hawaii Honolulu KFVE IND 5 Idaho Boise KBOI CBS 2.1 Idaho Idaho Falls/Pocatello KPVI NBC 6.1 Idaho Twin Falls KMVT FOX 14 Illinois Champaign/Springfield/Decatur WCFN MNTV 49 Illinois Chicago WCIU/WME IND 26.1/26.4 Indiana Evansville WEVV MyTV 44.2 Indiana South Bend WMYS IND 57 Kansas Wichita/Hutchinson KSCW CW 33.1 Kentucky

David Etherege

Lexington WDKY FOX 56.1 Kentucky Louisville WKYI IND 24 Louisiana Shreveport KMSS FOX 34.1 Maine Bangor WFVX Fox 22 Massachusetts Boston WBIN IND 50 Michigan Alpena WBKB IND 11.2 Minnesota Duluth KQDS FOX 21.1 Minnesota Mankato KETC FOX 12.2 Minnesota Minneapolis/St. Paul KSTC IND 5.2 Minnesota Rochester/Austin/Mason KXLT FOX 47.1 Mississippi Biloxi/Gulfport WXXV/WXX FOX 25.2 Mississippi Columbus/Tupelo/West Point WLOV

FOX 27.1 Mississippi Jasckson WRBJ CW 34 Missouri Springfield KCZ CW 33.2 Missouri St. Louis WRBU MNTV 46.1 Nebraska Lincoln/Hastings/Kearney KHAS NBC 5.1

Just a small list of Interviews secured by Shannon Rose

Anthony Amos

Guests

Getting guests for your show is simple invite them and they will be happy to be a guest it is that simple

Develop an audience with your show the guests will enjoy that even more as they interact with your social media marketing.

Always be marketing every show consistently for best results. The process of Consistently promoting creates momentum.

Updates will be added to the website and facebook page for ongoing education in the tech of podcasting.

Connecting with other podcasters via following other podcasts in your Niche or just supporting them by liking and sharing.

Connect with a local meetup FPA and attending Podfest.us for more connections I cannot emphasize enough the value of conferences like the Podfest Multimedia expo.

Every event equates to a new gold nugget that adds to your knowledge base to deploy to your podcast at every level new opportunities.

Insignia™ - Directional Microphone for Smartphone

Take notes and record thoughts with this Insignia unidirectional microphone. It connects to your phone using a 3.5mm jack for ease of operation, and it picks up your voice and isolates it so that you get clear, crisp sound. The wide frequency response ensures accurate sound reproduction from this Insignia unidirectional microphone

The News Format Podcast

HernandoPost.com
A Local Community is a great way to learn provide local content of community activity plus it is great fun to network with your home town.

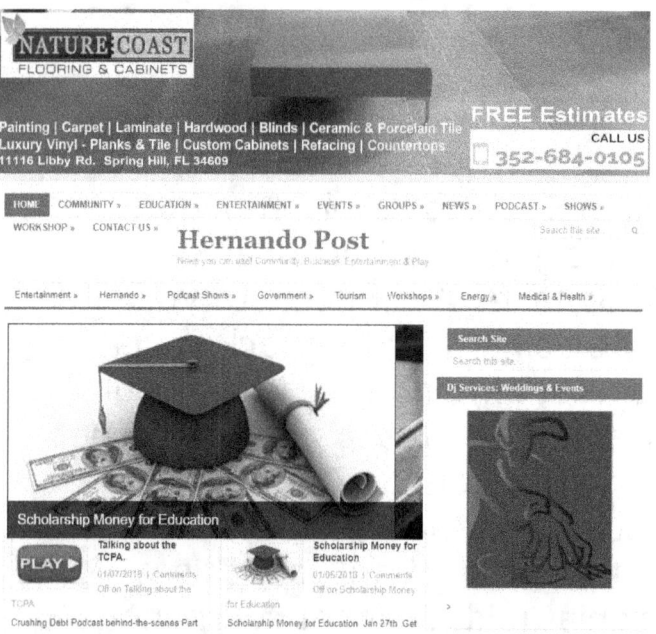

Need Help developing the systems of a local news magazine that will provide you with a steady marketing, guest and income stream.
All the Technology training done for you is available

TheTampaPost.com

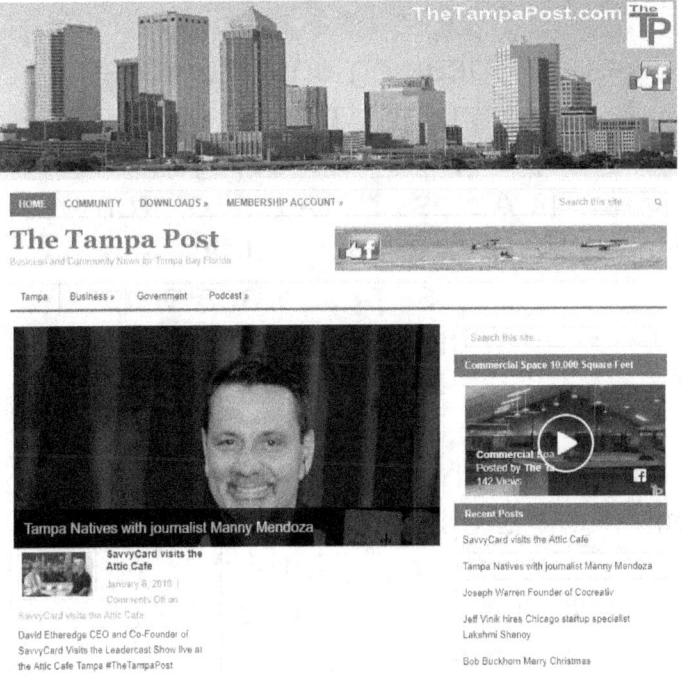

The Local news magazine blogs
have the following opportunities.

The Positive Business & Community New system.

1. Become the source of content in your local area
2. Invites to activities
3. Speaking opportunities
4. Fun and connections everywhere you go
5. Access to Deals to promote
6. Business to Politics it is all access pass when properly implementing the system

Most of the Podcast Technology in this book is used by anthony almost all the technology is interchangeable for multipurpose use for camera, smartphone or computer

Podcasting can take you any place you want to go!

www.ingramcontent.com/pod-product-compliance
Lightning Source LLC
Chambersburg PA
CBHW071219220526
45468CB00002B/677